ANIMAL TRICKS

BRIAN WILDSMITH

Oxford University Press
OXFORD NEW YORK TORONTO MELBOURNE

This is the tiger who stood upside-down.

And this is the leopard who's playing the clown.

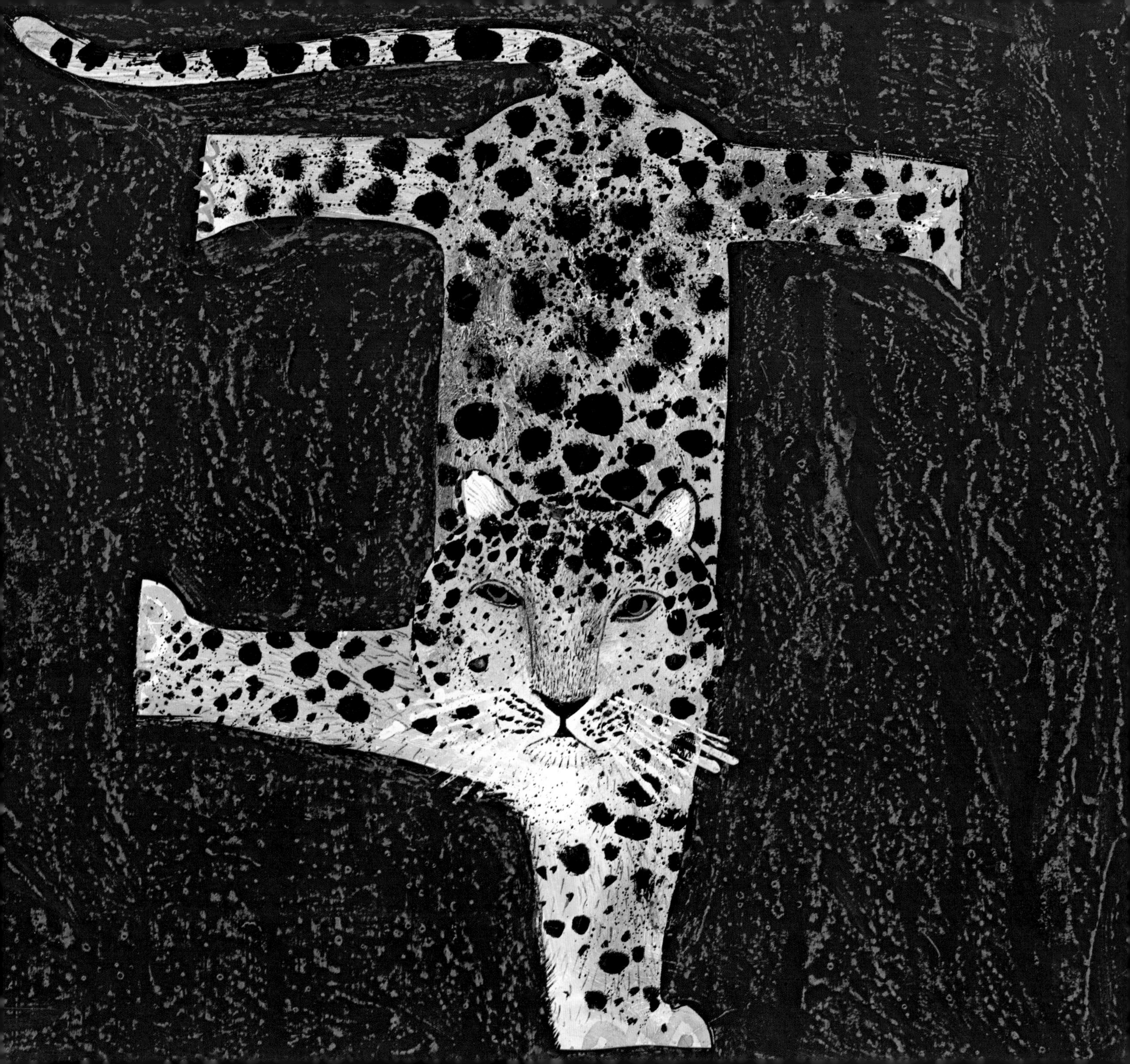

A pelican's beak can hold such a lot.

And here's a baboon with her tail in a knot.

A seal can balance a lot on his nose.

And a bear comes dancing on ten of his toes.

A brave little penguin who's really quite strong.

And look at the otter whose tail is so long.

The birds are nesting all over the moose.

And these two giraffes will never get loose.

Oxford University Press, Walton Street, Oxford OX2 6DP

Oxford is a trade mark of Oxford University Press

First published 1980
Reprinted 1983

First published in paperback 1991

ISBN 0 19 279743 3 (hardback)
ISBN 0 19 272176 3 (paperback)

Printed in Hong Kong